P9-CRJ-746

THE FACTS ABOUT FATS

A Consumer's Guide To Good Oils

John Finnegan
Elysian Arts

READER PLEASE NOTE: This book has been written and published solely for informational and educational purposes.

Please be advised that it is not the intention of this guide to provide medical advice or to substitute for the role of a physician in treating illness. The author and publisher provide this information with the understanding that you act on it at your own risk.

The Facts About Fats
A Consumer's Guide To Good Oils

All rights reserved. No part of this book may be reproduced or utilized in any form or by any means, electronic or mechanical, including photocopying, recording or by any information storage and retrieval system, without permission in writing from the publisher.

4A 2.00
CEC

© 1992 by John Finnegan
ISBN 0-927425-12-2

Published by

Elysian Arts
29169 Heathercliff Rd.
Suite 216-428
Malibu, CA 90265

This Book Is An Offering
To All People
Who Treaure And Care For
Life

Acknowledgements

I want to thank the many people who have given so warmly and freely of their time and help in producing this book. I especially want to thank my brother Todd for his enduring faith and support; Kathy Cituk who wrote some of the material, edited the text, helped design the cover and did most of the computer work, layout and design; and to Lynda Rayes, Francoise Paradis, Jeffrey Selkov, and Debra Pearlstein who showed up at the eleventh hour with their computer expertise, editorial skills, financial and moral suppport and midwifed the birth of this creation.

My special appreciation as well to Bob Walberg who has been a continual source of good humor, information, and honesty; Dr. Anine Lorenzen who drove me all over Europe and who tolerated (most of the time) my eccentric behavior; and Dr. Barbara Bieber who visited Dr. Budwig with me, spoke with the Kousmine Foundation, and translated their information and replies to my questions, which all contributed greatly to the comprehensive information contained within this book.

In this day and age of seemingly endless corruption in society, it is really refreshing to find people who have a genuine love and appreciation for life and who care about this earth and the creatures and plants living upon it.

I also want to express my deep thanks to Todd Chandler and Lisa Deladurantaye at Chandler Design, Mark Weideman at Premier Edible oils, Thomas Greither, Jorgen Uhrskov, Dr. Dan Roehm, Dr. Stephen Langer,

Zeigfreid Gursch, Udo Erasmus, Bill Vincent, Lucia Oswald, Robert Gaffney, Mrs. Rudin, Charlotte Gerson, John Hunt, Dr. Brian Roettger, Dr. Johanna Budwig, Danielle Fitzpatrick, Marlyese Ruess, Yvon Tremblay and many other friends and colleagues who assisted in bringing this book to fruition.

Thanks
For All Your Help
John Finnegan

How This Book Came To Be

In researching the material for this book I visited and examined the production facilities and processing methods of both Flora and Omega Nutrition in Ferndale, Washington, and Vancouver, Canada. I was graciously received and given a great deal of information and help by Bob Walberg, President of Omega Nutrition, and Thomas Greither, the owner of Flora. I also had extensive interviews and discussions with Robert Gaffney and Bill Vincent of Omega Nutrition and with Zeigfreid Gurshe, publisher of Alive magazine and former co-owner of Flora.

I flew to Germany and spent a most informative day in discussions with the charming and renowned Dr. Johanna Budwig. I next contacted the Kousmine Foundation and, although I was unable to visit the facilities and the Director, Dr. Kousmine, my colleague, Dr. Barbara Bieber, had extensive conversations with Dr. Kousmine and her staff, asking them questions and gathering research on my behalf.

I next visited Biogarden, one of Germany's largest natural foods distributors, and spent a day with their director, Rosie Reitschuster, discussing oil production and the quality of oils in the marketplace. Later I drove to Denmark and visited the organic, biodynamic farm of longtime environmental activist Jorgen Uhrskov. Jorgen is one of my oldest friends and is a real force for integrity and inspiration within the European environmental movement.

When I returned to the U.S., I sought out many of the leaders in the natural foods movement. I spoke at great length with the heads of some of the main companies producing health food oils, many of whom have a lifetime background in the organic foods and health food oil business.

I spoke with Mark Weideman at Premier Edible Oils and had extensive discussions with Frank Klann, owner of Galaxy, Ltd., and Ram Lakham Boodram, president of Seymour Organic Foods. They sent me photographs and diagrams explaining their oil processing methods.

I also contacted one of the nation's most well-known producers of health food oils, but they declined to discuss their oil processing methods with me and refused to allow me to visit their facilities.

I had several discussions with Udo Erasmus, author of *Fats and Oils* , and talks with Betty Kannen at OCIA, Hugo Skoppik at FVO, Christine Beeman at Allergy Resources, Clem Perrault who runs the largest organic farm in western Canada, Terry Cole, owner of Harmony Naturals which is an organic food distributor in New Zealand and Australia, and Charlotte Gerson, Director of the renowned Gerson Clinic. I also spoke with Dr. Stephen Langer, author of *Solved: The Riddle Of Illness,* Mrs. Rudin, the wife of Dr. Rudin, who is the co-author of the exemplary book *The Omega 3 Phenomenon,* and, finally, Dr. Dan Roehm who manages a cardiology clinic in southern Florida where he has personally witnessed the recovery of many patients from heart disease, cancer, obesity, and other diseases through the application of sound nutrition and the judicious use of flax seed oil and other therapies.

I have also seen first-hand the improvement and recovery from various diseases by many friends and clients who have implemented good living and dietary practices, proper use of oils, and herbal and nutritional supplements.

Finally, I spent thousands of hours studying the fats and oils literature, including some thirty books and one hundred-odd articles and scientific journals.

I hope this information will benefit many people. It took a lot to put it together. The fats and oils story may well be the greatest scandal of ignorance, disinformation, and greed in the entire history of food production. The effects of poorly processed oils are a major causative factor in heart disease, cancer, and most modern diseases that have affected hundreds of millions of people all over the world.

These are strong statements, and I urge the reader to examine the information for yourself and see what conclusions you reach.

John Finnegan

The Facts About Fats
A Consumer's Guide To Good Oils

Table of Contents

1

The Health Food Oil Scam

For years, even decades, I have gone into health food stores and bought my sesame, safflower and other oils with the assurance that I was purchasing good quality, health-promoting foods to use in my cooking and salad dressings. After all, I had read Paavo Airola and Bernard Jensen and had been informed by these and other respected authorities that commercial oils, which are sold in the supermarkets, are highly refined, nutritionally damaged by heat and full of poisonous chemicals. I am a conscious, intelligent person, taking responsibility for my health as well as trying to improve conditions on earth, so there's no way I would eat these polluted, hydrogenated carriers of evil carcinogenic chemicals.

Certainly, whatever food products I bought in the health food store would be good for me. Right? Imagine my shock when I discovered that our Emperor wears no clothes, that the public has been sold a bill of goods regarding the purity, safety and nutritional value of these so-called "health food" oils.

While the biggest hoax being perpetrated on the consumer public is that the margarines as well as Crisco,

Unico, Puritan, Mazola, and other vegetable oils are good for their hearts and health and that butter is bad; probably the biggest hoax being purveyed in the health food industry is the portrayal of the major producers of health food oils as suppliers of pure, unheated, chemical-free, nutritious cooking and salad oils.

To be sure, we do have these companies to thank for providing us with expeller-pressed (albeit at temperatures of 180 to 210 degrees F. and even much higher) instead of solvent-extracted oils, but still most of these oils are refined.

What is wrong with refined oils? Refined oils are subjected to several processing methods— deodorization, winterizing, bleaching, and alkali refining.[1,2,3] These processes remove virtually all the vitamin E, lecithin, and beta carotene.[4] Even worse, refining destroys the essential Omega 3 and Omega 6 fatty acids, converting them into poisonous trans fatty acids.[5,6] The oils also develop a certain amount of rancidity because they are processed in the presence of light and oxygen, then bottled in clear glass containers which allow the light to penetrate and further their rancid deterioration.[7,8] Light causes serious free radical damage to oils, and light oxidation is 1,000 times faster than oxygen oxidation.[9]

The deodorizing phase of the refining process is the worst. The oil is heated to temperatures from 470 to 518 degrees F. which removes the strong taste of the seed from which it is extracted.[10] Why do all these companies make such a big deal about their oils being extracted at a low temperature (which even at 180 to 210 degrees F. is still high enough to cause some damage and ideally should be below 118 degrees F.) and not inform us that they also subject the oils to a 470 to 518 degree temperature in the deodorizing process?

Why is it necessary to deodorize the oils anyway? I like the taste of the seeds. Almost all the oils used in Australia and many of the oils used in Europe are unrefined. Are we Americans and Canadians so obsessed with blandness that we insist on having oils like these, or is this something that is being perpetrated on an unknowing and mostly uncaring public?

To add further insult to the injury already done to oils, they are packaged and distributed in clear glass bottles which let light in, causing the oils to become rancid. Some companies do make truly fresh-pressed, unrefined oils and package them in dark glass bottles, which is a great improvement; however, there is real evidence that some damaging light rays do penetrate the dark amber glass bottles and cause a certain amount of rancidity.[11,12,13,14] Metal does not make a good container for oil either because it will contaminate the oil. The best solution is to package the oil in either an opaque glass or a special black plastic container that keeps virtually all light out and has been proven to allow no transmigration of hydrocarbons from the plastic into the oil.[15,16,17] The only companies that are currently providing truly cold-pressed, organic, unrefined oils in such containers are Omega Nutrition, Omega/ Arrowhead Mills and Galaxy Enterprises.

Why are we being sold oils that purport to be healthy and nourishing and yet in reality are refined, devitalized and filled with poisons? Mostly because we accept it. We trust these companies, and we do not know the truth.

If we want foods that are healthy and non-carcinogenic, it is up to us to educate ourselves, request that our stores carry the good brands, and then support them with our buying power even if these foods cost slightly more.

Just as we need good quality protein, carbohydrates, vitamins and minerals, we need good quality fats and oils to be physically and mentally healthy. There are two types of fats—saturated and unsaturated. The saturated fats are found in butter, eggs, fish, chicken and meats, and are high in cholesterol. The unsaturated fats are found in vegetable oils such as sunflower, sesame, safflower, corn, and flax seed oil, and are high in the essential Omega 6 and Omega 3 fatty acids. The non-essential monosaturated fatty acids are found in olive and other oils.

We need a certain amount of both saturated and unsaturated fats in our diets. Except for people who have serious heart disease from a lifetime of poor eating and little exercise, most of us benefit from a minimal amount of cholesterol in our diet. Over eight percent of our brain's solid matter is made of cholesterol; our hormones, skin, and even the membranes of our cells use cholesterol as an essential building block in their basic production and structure.[18,19]

Contrary to media disinformation, butter is an excellent food. Butter is a very different fat from the fat in animal tissue. It is high in butyric acid which is an excellent source of energy, and it is a good source of cholesterol, which most of us benefit from. Even most people with dairy allergies tolerate butter well, because usually they are allergic to the lactose (milk sugar) and milk protein, not the milk fat. Unfortunately, butter today is not as good as it could be, since most cows are fed grain which is grown with pesticides, thus passing along a trace amount of those poisons in their milk. However, butter is still an excellent food, except for people with high cholesterol levels. Maybe soon we will be able to obtain butter from cows that are fed organically-grown grains. That would be nice.

We also need the unsaturated fats—the Omega 6 and Omega 3 fatty acids—found in the vegetable oils. It isn't a question of either/or—should I use saturated or unsaturated fats? Most of us need both in good-quality forms. What we don't need are the refined or hydrogenated fats found in margarines, vegetable shortenings or refined vegetable oils. Excellent information has been brought forth in recent years by Dr. Johanna Budwig, Charles Bates, Ph.D., Dr. Donald Rudin, and others, showing the critical need for the Omega 6 and Omega 3 fatty acids in our diets. These fats are good sources of energy; they are an essential part of the body's oxygen transport mechanism; and they are key building blocks of our skin, hormones, etc. Most of us obtain ample amounts of the Omega 6 fatty acids from safflower, sesame, sunflower, soy, canola and corn oils, but we are seriously deficient in the Omega 3 fatty acids.

A recently-published collection of research papers presented at the NATO Advanced Research Workshop on Dietary Omega 3 and Omega 6 Fatty Acids determined that, for normal usage, the correct ratio of Omega 6 to Omega 3 fatty acids is 9 to 1. Other authorities feel the ratio should be 6 or 4 to 1, which is approximately 2-4 tablespoons a day of polyunsaturated vegetable oils (safflower, sunflower, etc.) and 2-3 teaspoons a day of flax seed oil.[20]

Obviously, these are just general guidelines, as everyone has individual needs, and some people's metabolisms burn much higher amounts of fat than others. Also, people who have developed heart disease or other illnesses from a lifetime of excessive consumption of poor quality fats and a deficiency of the Omega 3 fats, will initially need higher therapeutic doses of Omega 3 fatty acids.

A deficiency of Omega 3's has been strongly implicated as a main cause of heart disease, cancer, immune system breakdown and other modern maladies.[21,22] The only really concentrated sources of the Omega 3 fatty acids are flax seed oil and certain fish, such as mackerel, sardines, tuna, trout and cold water salmon. This is why we need one or both of these sources in our diets. While eating these fish can be a good source of the essential fatty acids, the use of fish oil extracts in capsule and other forms may not be the best thing for us. According to research done by Dr. Johanna Budwig and others, using fresh-pressed flax seed oil is far superior to using fish oil extracts because the processing methods for fish oils cause real damage to the good fats and also create many toxic compounds.[23]

Nearly all the oils distributed by our major suppliers of health food oils are refined and bottled in clear glass containers. This means that most of the nutritional content of the oils has been removed, that they have been exposed to high heat (470 to 518 degrees F.) in the deodorizing process, that the nourishing cis fatty acids have been converted into poisonous trans fatty acids, and that they are developing rancidity from being exposed to air and light in their processing and further deterioration from light while stored in clear glass bottles. Sounds like just the kind of food we want to eat.

As a matter of fact, most of these so-called health food oils are made by the very same companies, using most of the same processing methods and equipment that produce the oils for Unico, Mazola, Puritan, Wesson, etc. These companies simply bottle the oil that is crushed and refined by other companies like Liberty Vegetable Oils, Elders, Producers, etc., the very same companies that make the oils for the supermarket brands. One of the main health food oil companies did actually begin producing their own flax seed oil, but they lost their FVO Organic certification in 1989 and they are now certifying themselves to be organic, which is basically meaningless.

In the past several years, a number of health food companies have begun bottling unrefined oils. In some ways these appear to be a great improvement, but they still have real drawbacks. These oils develop rancidity because they are processed in the presence of light and oxygen, and, worst of all, they are bottled in clear glass containers which allow light to enter and create free radicals in the oils. Another major concern is that unrefined oils from non-organic grains can have high levels of pesticides and herbicides. Also, most expeller-pressed oils are processed at temperatures of 140 to 160 degrees F. or even higher.

Good-quality oils need to be expeller-pressed at temperatures below 118 degrees F., not solvent extracted. They should not be subjected to high heat temperatures, to deodorizing, bleaching, alkali refining, or winterizing processes. They need to be produced by light and oxygen excluding methods, bottled in containers that prevent the further exposure to light which causes rancidity, and, where possible, they should be produced from organically grown seed.

At this time, I have found only four companies in Canada and the United States producing oils which meet these standards of quality: Seymour Organic Foods, Flora, Inc., Galaxy Enterprises, and Omega Nutrition. Arrowhead Mills distributes the full line of Omega Nutrition oils in the U.S.; the Omega Nutrition/Arrowhead Mills and Galaxy Enterprises oils have the added advantage of being the only oils bottled in completely light-excluding containers. Omega Nutrition/Arrowhead Mills are the only companies producing a full line of organic fresh pressed vegetable oils that have both FVO and OCIA (Farm Verified Organic and Organic Crop Improvement Organization) certification. This is a crucial concern today because many companies are falsely representing their products to be organic when in reality they are not. Without independent third party certification from a reputable company, the consumer has no guarantee of the quality of the products they purchase.[24]

1. Rudin, Donald O., M.D., and Felix, Clara,*The Omega 3 Phenomenon,* New York: Rawson Associates,1987.
2. Swern, Daniel, Editor,*Bailey's Industrial Oil And Fat Products,* New York: John Wiley and Sons,1979.
3. Gittleman, Ann Louise, M.S., *Beyond Pritikin,* New York: Bantam,1989.
4. Rudin, op cit.
5. Gittleman, op cit.
6. Rudin, op cit.
7. Gunstone, Harwood, and Padly, *The Lipid Handbook,* London: Chapman and Hall,1986.
8. Gittleman, op cit.
9. Gunstone, et al., op cit.
10. Swern, op cit.
11. Gunstone, et al., op cit.
12. Carlson, D.J., Supruchuk, T., and Wiles, D.M., "Photo-oxidation of Unsaturated Oils: Effects of Singlet Oxygen Quenchers," Division of Chemistry, Nat. Res. Conl. of Canada, J.O.A.C.S. Vol. 53, October 1976, p. 656-9.
13. Faria, J.A.F., de Vicosa, U., and Mukai, M.K., "Use of a Gas Chromatographic Reactor to Study Lipid Photo-oxidation," Rutgers Univ., New Jersey, J.O.A.C.S. Vol. 60, No. 1,1983.
14. Recent study at the University of British Columbia facilities by Pete Vincent, Engineering Research Assistant, Technologist/Physicist.
15. *Journal of American Oil Chemists Society,* Vol. 62, Aug. 1985, Study of Oxidation by Chemiluminescence. IV. Detection of Levels of Lipid Hydroperoxides by Chemiluminescence. Y. Yamamoto, E. Niki, R. Tanimura, Y. Kamiya, Dept. Reac. Chem. Fac. Engr., U. of Tokyo, Japan.

16. Ibid., Vol. 61, Mar.1984, Flavor and Oxidative Stability of Hydrogenated and Unhydrogenated Soybean Oils. Efficacy of Plastic Packaging. K. Warner, T.L. Mounts, N.R.R.C., Agr. Research Ser., USDA.
17. Independent testing done by Cantest, Vancouver, B.C., showing no transmigration of hydrocarbons from black plastic used by Omega Nutrition.
18. Gittleman, op cit.
19. *East/West Journal,* February 1988, "Butter vs. Oil" by Rudolph Ballentine, M.D.
20. Galli, Claudio, and Artemis P. Simopoulos, *Dietary Omega 3 and Omega 6 Fatty Acids: Biological Effects and Nutritional Essentiality,* New York and London: Plenum Press, 1988.
21. Budwig, Dr. Johanna, *Flax Oil As A True Aid Against Arthritis, Heart Infarction, Cancer and Other Diseases,* Vancouver, Canada: Apple Publishing, 1992.
22. Erasmus, Udo, *Fats and Oils,* Canada: Alive, 1987.
23. Budwig, op cit.
24. Carlat, Theodore Wood, *Organically Grown Food,* California:Wood Publishing 1990.

2
Causes of Essential Fatty Acid Deficiency

The dramatic changes in our agricultural, food processing, food preparation methods and dietary habits that have occured over the past one hundred years have brought about an unprecedented alteration in people's nourishment with serious consequences effecting the widespread increase in degenerative illneses in nearly all technologically developed societies.

Following are the major changes causing the development of this widespread deficiency of the essential fatty acids in mankind today.

1. Change in flour milling technology, causing elimination and rancidity of essential fatty acids.

2. Elimination of Omega 3 foods, such as flax oil, because of limited shelf life. Producers want products that last for years on the supermarket shelf. Introduction of refined and hydrogenated oils which are high in poisonous trans fatty acids, rancid fats, and free radicals and deficient in vital Omega 3 and Omega 6 fatty acids. The average western person today consumes 1000 percent more trans fatty acids and hydrogenated fats than before.[1]

3. Change to feed lot-raised beef as a primary protein source along with caged chickens and eggs. Change from wild game, range-grown beef, deer, turkeys, and sheep, fish, etc., to cage-raised beef, chicken, eggs and dairy products as main protein sources. Free range-grown beef, chicken, eggs and dairy products can have up to 5 times as much Omega 3 and Omega 6 fats in their tissues and a much lower amount of hard cholesterol fats as their caged counterparts.[2] The same change has happened to farm-raised rainbow trout and salmon which normally have a high content of essential fatty acids and their derivitives. When grown on fish farms their essential fatty acid content is substantially reduced because their normal foods like algae which are high in the EFA's are replaced with soy meal and other less nutritious foods.

4. Certain races have an inherited need for more essential fatty acids and GLA resulting in a much higher need for these nutrients in the diet. The genetic heritage of these population groups – Celtic Irish, Scottish, Welsh, Scandinavian, Danish, B.C. Coastal Indians, and Eskimos – leads them to need more EFAs in their diets, because their ancestors lived on large amounts of fish high in these nutrients. They are much more prone to develop deficiency diseases when their diets lack sufficient quantities of these key nutrients.[3,4]

5. Increased use of drugs and pharmaceuticals, particularly aspirin, that block EFA enzyme systems and their conversion to vital prostaglandins.

6. Increased use of sugar, caffeine, refined carbohydrates, and alcohol which deplete EFAs and prostaglandins. Alcohol and caffeine also block conversion of EFAs to prostaglandins.

7. Increased ingestion of toxins in food, water, and air which deplete EFAs.

8. Lack of breast feeding. Omega 3 fats and DHA are not presently found in infant formulas or commercial cow's milk. They are also deficient in the breast milk of mothers whose diets are deficient in Omega 3 fatty acids.

9. Excessive consumption of Omega 6 fatty acids, which interferes with the absorption of Omega 3 fatty acids.

---

1. Rudin, Donald O., M.D., *The Omega 3 Phenomenon,* New York: Rawson Associates,1987.
2. Rudin, Ibid.
3. Rudin, Ibid.
4. Bates, Charles, Ph.D., *Essential Fatty Acids and Immunity in Mental Health,* Washington: Life Sciences Press, 1987.

How Oils are Manufactured

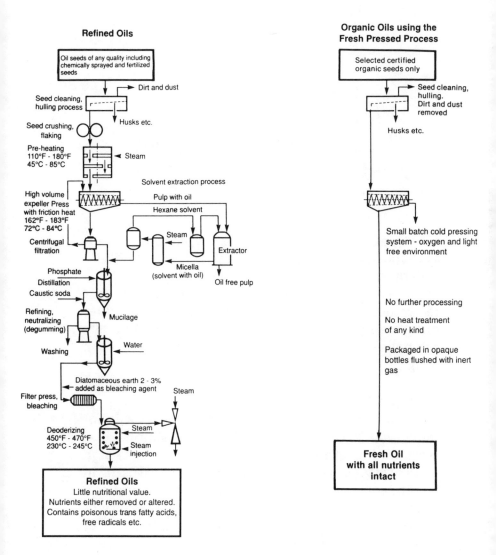

Refined Oils

Oil seeds of any quality including chemically sprayed and fertilized seeds

Seed cleaning, hulling process → Dirt and dust

Husks etc.

Seed crushing, flaking

Pre-heating
110°F - 180°F
45°C - 85°C ← Steam

High volume expeller Press with friction heat
162°F - 183°F
72°C - 84°C

Solvent extraction process

Pulp with oil

Hexane solvent

Centrifugal filtration ← Steam

Extractor

Micella (solvent with oil)

Oil free pulp

Phosphate
Distillation
Caustic soda

Refining, neutralizing (degumming) → Mucilage

Washing ← Water

Diatomaceous earth 2 - 3% added as bleaching agent

Filter press, bleaching → Steam

Deoderizing
450°F - 470°F
230°C - 245°C ← Steam ← Steam injection

Refined Oils
Little nutritional value.
Nutrients either removed or altered.
Contains poisonous trans fatty acids,
free radicals etc.

Organic Oils using the Fresh Pressed Process

Selected certified organic seeds only

→ Seed cleaning, hulling. Dirt and dust removed

Husks etc.

Small batch cold pressing system - oxygen and light free environment

No further processing

No heat treatment of any kind

Packaged in opaque bottles flushed with inert gas

Fresh Oil with all nutrients intact

© 1992 John Finnegan

3
Vegetable Oils:
Their History, Properties
And Uses

Questions often arise such as: "Which oil is best for sautéing? For baking? For salads? For mayonnaise?" Here are some simple guidelines. Generally, all unrefined oils have a lower smoke point, and, therefore, are not recommended for high temperature frying. Both butter and ghee are included in this section, because while they are not vegetable oils, they are excellent fats to cook with.

Almond Oil
The almond tree has been cultivated in Asia for more than 35 centuries. Unrefined sweet almond oil is sweet and pleasant tasting and is known for its high content of vitamins A & E. It has long been used as a skin beautifier, as an antiseptic for the intestines, and as a source of minerals. Therapeutically, almond oil has been used in treating gastric ulcers and as a laxative, as well as to help stabilize the nervous system. Food-grade

almonds are expensive nuts, making a quality almond oil expensive and very difficult to find in a truly cold-pressed form. Therefore, it is mostly used topically as a massage oil or for treatment of burns.

Borage Seed Oil

Borage seed oil was first used during the Middle Ages when it was used to insure good quality of blood. Later, during the Renaissance, it was recommended to cure depression and to strengthen the heart. Borage seed oil is high in gamma-linolenic acid, a derivative of the Omega 6 fatty acids, and has been used to help those struggling with illnesses including arthritis, allergies, multiple sclerosis, cancer, arteriosclerosis, immune system deficiency, diabetes, obesity, PMS, alcoholism, liver degeneration, colitis, depression, skin disorders, and other conditions. Borage seed oil should only be used in small amounts, (1/4 to1/2 teaspoon daily). It can be taken plain by mouth or poured directly onto food. It should never be used in cooking and should be kept refrigerated. One can also take it in capsules, and some companies are making a flax-borage seed oil combination capsule that provides an excellent balance of essential fatty acids with gamma-linolenic acid. Borage oil contains twice as much GLA (gamma-linolenic acid) as Evening Primrose oil.

Butter

Butter has been around for thousands of years and can be used for any type of cooking. It is stable in the presence of light, heat, and oxygen. Nothing compares to the taste of butter; however, it is not high in essential fatty acids so it should not be considered a source for these essential nutrients. In moderation, the cholesterol content of butter should not be a problem as long as the diet has sufficient essential fatty acids, is free of poisonous fats, and contains those elements necessary for the efficient metabolism of fat.

Canola Oil

Canola is a hybrid of two species, a cousin of the cabbage, and dates back to antiquity in Eastern Europe and Asia. Recent strains of the plant contain less erucic acid than their ancestors, and the oil is gaining in popularity. Canola oil is low in saturated fats and high in monounsaturated fats. Therapeutically, it has been used to protect artery walls and as protection against blood clots. It is a delicate oil with a fresh, golden color and is best used directly on salads, raw vegetables, potatoes, grains or cottage cheese. It has a fine, savory flavor which is excellent in salad dressings and mayonnaise. Due to its Omega 3 fatty acid content, canola oil should not be used for cooking over 320 degrees F. Unrefined canola oil has a rich, strong flavor and is wonderful as a fat replacement in bread recipes. I find both flax and canola to be especially delicious, use them on all my dishes, and often dip my bread or crackers in the oils instead of using butter.

Evening Primrose Seed Oil

Evening Primrose is similar to Borage seed oil in that it has a high content of gammalinolenic acid. It has also been used to help people overcome PMS, allergies, arthritis, depression, skin disorders, colitis, multiple sclerosis, liver degeneration, alcoholism, cancer, arteriosclerosis, immune system weakness, obesity and other disorders. Some tribes of American Indians used the herb for asthma, skin disorders and other conditions. Evening primrose oil is available in both liquid form and capsules.

Ghee

Yvon Tremblay, a renowned French Chef from Montreal, wrote the following statement about his use and appreciation of ghee.

"One of my favorite ways of cooking is to mix 1/2 ghee with 1/2 oil in preparing, marinating or sautéing vegetables, fish or meat. The taste is without compare and the ghee will never burn if it is prepared properly.

The smell of the ghee and oil cooking is exquisite and will have such an effect on your guests that they will rave to find out what has given your dish that special taste. This is one of the most treasured secrets of fine chefs."

There are many reasons why the best chefs prefer ghee and the ghee-oil mixture for their cuisine.

- Ghee will never burn the dish you cook with.
- It doesn't smoke like butter and develop toxic compounds if overheated.
- The aroma is completely different in one's cooking.
- Pastry chefs love to coat their pastries with ghee.
- Ghee is the most highly regarded cooking fat by two of the finest culinary arts in the world, the French and the Indian.

Prepared ghee can be purchased in most health food stores and super markets. One can also make it oneself by using the following method:

- Take 2 pounds of unsalted butter, cut into pieces and place in a cooking pot.
- Melt the butter using a low temperature until it is liquid and you can see the fat on the top bubbling.
- Cook on a low flame for 20 minutes.

The impurities will rise to the top.

- Place a piece of cheese cloth in a colander and strain the butter, or simply scoop the light-colored impurities and milk compounds from the top of the melted butter.

Ghee will keep much longer than butter and will store well even without refrigeration.

Hazelnut Oil

The hazelnut is believed to have its origins in ancient Greece. Hazelnut oil has long been popular in the finer European kitchens. The delicate flavor and bouquet of the hazelnut is exceptional on salads and pasta as well as in pancakes, waffles, and muffins. Hazelnut oil is rich in monounsaturated fatty acids, similar to olive oil. This oil is considered to be the finest gourmet cooking oil, preferred by many Gold Medal chefs for their use in creating fine dishes and desserts. It is also an excellent oil to massage into the skin. Traditionally, it has been used in the treatment of parasites, tuberculosis, urinary disease, and colitis. It is recommended to people recovering from disease, the elderly, pregnant women, and diabetics. It is an easily digested oil.

Olive Oil

In Greek mythology, Poseidon and Athena were fighting over the most beautiful area in Greece. To avoid a conflict, it was decided that the humans would choose their protecting god for themselves. To win their favor, Poseidon produced a marvelous horse. Athena brought a tree to the people— an olive tree. She won, and Athens was built in her honor. Fresh olive oil has a wonderful, rich, aromatic bouquet. It has been used therapeutically to nourish the liver and gall bladder, and is recommended for use by diabetics. Olive oil is high in monounsaturated fatty acids and is helpful in reducing cholesterol. It has also been used topically for soothing burns, eczema, and psoriasis. Olive oil has been used for centuries in Greek, French, and Italian cuisines, and is a favorite for giving even ordinary dishes a "Mediterranean" flavor. It is excellent for sautéing, makes a rich, exotic mayonnaise, and is delicious when used in salad dressings, sauces, and spreads.

Pistachio

Pistachio nut oil is an exquisite green oil with a rich, sweet taste. It has the finest taste of all the gourmet cooking oils and is highly prized for the special flavor it lends to fine dishes.

Pumpkin Seed Oil

Pumpkin seed or squash seed oil has been used throughout history in India, Europe and the Americas. It has a good proportion of both the Omega 6 and Omega 3 essential fatty acids making it among the most nutritious of oils. It is quite tasty and should be used raw, poured directly onto vegetables, pasta, and other dishes. It has similar properties to the seeds and has been used to nourish and heal the digestive tract, to fight parasites, to improve circulation, to help heal prostate disorders, and to help prevent dental caries. It is also recommended to pregnant and lactating women for its essential fatty acid content. A lack of certified organic pumpkin seeds has prevented the large-scale production of this oil by the few reputable oil companies.

Safflower Oil

The safflower, a member of the daisy family, is a native of the mountainous regions of southwest Asia and Ethiopia, and is grown extensively in India. This annual plant yields flowers, which are used in dye, and oil-rich seeds. Safflower oil is probably the most versatile vegetable oil, and, in its unrefined form, if manufactured correctly, has a unique wonderful flavor. It can be used for sautéing, baking, stir-frying, and it makes excellent dressings, sauces, dips, and mayonnaise. It is particularly tasty in grain salads. Safflower oil is rich in polyunsaturated fatty acids.

Sesame Oil

Sesame oil is the traditional oriental and macrobiotic cooking oil, providing the familiar sesame flavor in sautéed dishes. The sesame seed constitutes a mainstay of nutrition in the Middle East, especially in Turkey. It is rich in lecithin, which helps build the nervous system and brain cells. It has been used to help depression and stress as well as to improve circulation. Sesame oil is rich in monounsaturated and polyunsaturated fatty acids, is very versatile, and can be used for all cooking needs. It also contains sesamol, a natural preservative, so it is very stable. Sesame oil is a delicious ingredient for mayonnaise, dressings, spreads, and pasta toppings. It is also an excellent massage oil.

Sunflower Oil

The sunflower originated in South America, and the Incas, who worshipped the sun as their god, used its oil in many ways. It has a long history of being used to help the endocrine and nervous systems, and it reduces cholesterol levels. Sunflower oil will add the delicate nutty flavor of fresh sunflower seeds to salads, baked goods, and other dishes. It can be used for baking and sautéing, and it adds a wonderful flavor to salad dressings, sauces, and dips. It is rich in polyunsaturated fatty acids and vitamin E.

Walnut Oil

Walnut oil has a delicious flavor, is great to pour on salads or other dishes, and certain varieties of walnuts are an excellent source of the Omega 3 and Omega 6 fatty acids. However, since all the walnut oils currently available in stores are heavily processed and refined, they are not good to use. Omega Nutrition will manufacture fresh- pressed oil from organic walnuts for special orders.

Several oils are not included in this section for the following reasons:

Cottonseed Oil

There currently is very little organic cottonseed available. Cotton is one of the most heavily sprayed crops. Cottonseed has 48% Omega 6 (Linoleic), 23% Omega 9 (Oleic acid), and 23% Palmitic acid.

Corn Oil

Because of the low oil content in corn, extremely high temperatures and toxic solvents are needed to extract the oil efficiently. Corn oil has 49% Omega 9 (Oleic acid), 34% Omega 6 (Linoleic acid) and 11% Palmitic acid.

Soy Oil

It is especially difficult to extract oil from soybeans. They need to be roasted and treated by other means that damage the oil, and they also are subject to further high temperatures and toxic solvents in the extraction processes. Soybean oil has 51% Omega 6 (Linoleic acid), 29% Omega 9 (Oleic acid), 10% Palmitic acid and 7% Omega 3 (Linolenic acid).

Tropical Oils

Palm oil, palm nut and coconut oils do not contain either of the essential omega 6 or omega 3 fatty acids to any appreciable degree, (with the exception of palm oil which contains 10% omega 6). Their fats are easily digestible and function mostly as food for the body to use in energy production.

A few years ago there was a disinformation campaign excerted through the media to scare the American consumer away from "tropical oils" saying that their fat content contributed to heart disease and that businesses and consumers should replace them with hydrogenated vegetable oils like soybean oil. This is another example of the results of well meaning but misinformed spokesmen for our health.

Dr. Everett Coop former Surgeon General of the United States termed the tropical oil scare "Foolishness" and continued to state "but to get the word to commercial interests terrorizing the public about nothing is another matter".

As I have documented in this book, the problems of heart disease are many fold and actually poisonous trans fats that are produced when unsaturated fats are subjected to refining and hydrogenation processes are much more implicated in causing arteriosclerosis than good quality saturated vegetable fats. [1]

And there have been studies that have shown that good quality saturated fats from vegetable oils, (all of which contain no cholesterol) actually help to lower cholesterol and prevent heart disease.

Palm Oil

Palm oil has been used for more than 5,000 years, produced from the fleshy part of the palm fruit. It is the second most produced oil in the world today. Annual production being 10.33 million tons, with soybean oil being the largest at 15.04 million tons out of a total annual vegetable oil production of 77.33 million tons (in 1989).

Palm oil is a very interesting oil with many unique and beneficial properties. Palm oil is extracted solely by mechanical and physical methods without the use of poisonous solvents like hexane that are commonly used to process other oils. Palm oil in its crude form is among the richest sources of beta-carotene and has a high content of Vitamine E. Crude palm oil contains 300 times more beta-carotene than tomatoes and 15 times more beta-carotene than carrots.

However since all palm oil used in the western world is refined (crude palm oil has a rich reddish brown color which most consumers would probably reject) as usual most of the vital nutients are destroyed. The introduction of unrefined, organically grown palm oil packaged in opaque containers to the western countries would be a valuable and nutritious product.

Palm oil has a good fatty acid profile containing 40% omega 9's (oleic acid-monosaturated), 44% palmitic acid (saturated), 10% omega 6's (linoleic acid-unsaturated) and 5% stearic acid (saturated).

There are many characteristics making palm oil desireable for use in manufactured food products. It's fatty acid composition gives it a desired semi solid consistency without needing hydrogenation, it is very resistent to oxidation and therefore has a good shelf life, holds up well in hot climates and has properties making

23

it excellent to use in cakes and bakery products. It is also readily available and is inexpensive to produce relative to most other oils.

It is not curently recommended for use because only refined palm oil is available in North America and Europe. However, it is better to use than other heavily refined hydrogenated vegetable oils.

Palm Kernel Oil

Palm kernel oil is extracted from the nut of the palm fruit. It is 51% saturated fat (lauric acid), 18% unsaturated (myristic acid), 9% saturated (palmitic acid) and 14% (stearic acid). It's main advantage over other commercial oils is the same as palm oil in that it isn't subject to hydrogenation in its processing methods making it a purer oil.

Coconut Oil

Coconut oil is composed of 90% of the medium chain triglyceride saturated fats. It is made up of 48% lauric acid, 17% myristic, 9% palmitic, 8% caprylic, 7% capric, 6% oleic and a small percentage of several other fats. Coconut oil is excellent for cooking because of its heat resistent properties. It's fats are easily digested and metabolized and don't tend to cause weight gain. Besides food uses, it is also the primary fat used in soap making. Again if it were only available in an organically grown, unrefined state and packaged in light excluding containers we would have an excellent fat for cooking purposes.

Cooking Methods

You may have noticed that we do not recommend any oil for use in deep frying. The intense heat required for deep frying will destroy some of the properties of any oil as well as create poisonous compounds and toxic fatty acids; therefore, this method of cooking is not recommended.

That does not mean, however that oils should be eliminated altogether in cooking. Flavor and nutrition content can be enhanced with the proper use of oils. Wonderful combinations of vegetables, herbs and spices, with tofu, nuts or seeds added for variety, can be lightly sauteed in olive, hazelnut, sesame, safflower, or sunflower oil.

- Place one to two tablespoons of oil in a frying pan, skillet, or wok. Heat the oil slowly over a low flame, then add ingredients according to the length of time needed to cook. (Denser, harder vegetables should go in first. Then add the lighter vegetables, with leafy vegetables last.) Stir often. One can also add an equal amount of butter or ghee to the pan to make the delicious oil-butter or oil-ghee mixture.

- If vegetables start to stick, add a quarter to a half cup of water, broth, or soup stock, and cover the pan. The steam from the liquid will cause the vegetables to cook quickly.

- To turn the broth into a thick, delicious sauce, dissolve a teaspoon of arrowroot or kudzu in a small amount of cold water. Stir the mixture into the vegetables about a minute before they are finished.

---

1. Mensink and Katan, "Trans Fatty Acids and Lipoprotein Levels" *New England Journal of Medicine*, Vol. 323, NO. 7, August 16, 1990.

4

The Wonder of Flax Seed Oil

Flax is a very ancient plant. It was used and highly prized for its nourishing and healing properties by many civilizations throughout history. Europeans, American Indians, ancient Egyptians and other cultures used flax seed oil. Up until the late 1800's and early 1900's, with the development of high technology and the mass production of refined oils, many towns and cities in Europe and North America still had small family-run workshops where they produced freshly made flax seed and other oils and delivered them door to door just like the farmers bringing around fresh milk.

The Cherokee Indians regarded flax as one of their most nourishing and healing herbs. They believed that flax seed oil captured energies from the sun that could then be released and utilized in the body's metabolic processes. This humble little plant was as sacred to them as the eagle feather. They beat the seeds in a gourd until the golden oil was freed from its shells and dripped gently into a waiting bowl. Then they mixed this oil with curdled goat or moose milk (to form special lipoprotein compounds), honey and cooked pumpkin. They used the oil by itself as well as the oil, protein, honey and pumpkin dish to nourish pregnant and nursing mothers and give them needed nutrients for creating strong and healthy children. They also fed this dish to people who had skin diseases, arthritis and malnutrition, and to men to increase their virility.

Dr. Johanna Budwig and other doctors also feel that mixing or blending flax seed oil into good protein like raw, cultured, lowfat cottage cheese will nourish the body better. She has repeatedly observed that the flax seed oil and protein combination form special lipoprotein compounds that are easily digested, and the body will use them to build new tissues with. She, Dr. Kousmine and other doctors in Europe use flax seed and other vegetable oils poured on dishes and mixed with non-fat yogurt and/or cottage cheese as an essential part of their succcessful dietary therapies for cancer, heart disease, arteriosclerosis, arthritis and many other modern maladies.[1,2,3] People who have dairy allergies can use tofu instead, or a high quality protein formula like the Ultra Clear or Ultra Balance formulas distributed by Metagenics/Ethical Nutrients. Further information and specific recipies using flax seed oil and the flax seed oil protein combinations are contained in my book *Fats And Oils: Promise Or Poison.*

Dr. Rudin and other authorities feel that a deficiency of the Omega 3 fat has created widespread illness in the population similar to other conditions that previously existed like scurvy (caused by a deficiency of vitamin C) and pellagra (caused by a deficiency of vitamin B3).[4,5,6] He and other doctors have successfully treated schizophrenia, depression and other emotional disorders by using a good diet supplemented with 1 to 2 tablespoons a day of flax seed oil daily.[7,8]

Depression is without question the foremost emotional disorder today and Dr. Rudin feels, with good reason, that a deficiency of the Omega 3 essential fatty acid is a major cause of this condition. While it is clear that people's estrangement from their true self is the greatest cause of unhappiness today, there is no question that good nutrition also plays an essential role in our emotional health and well being.

Besides being the most inexpensive and practical source for the Omega 3 fats, flax seed oil is an excellent source of the Omega 6 fats, carotene and vitamin E.

There have been many scientific studies demonstrating the healing properties of flax seed oil:

A research project in Australia used flax seed oil and linolenic acid to successfully fight strep infections in hospitals in Victoria.[9]

A study done in Poland observed that "Fatty acids isolated from linseed oil were found to exhibit a strong cytotoxic (cancer cell destroying) in vitro (outside the body) activity, against (certain) cancer cells with minimal cytotoxic effect on normal cells (leukocytes) of rabbits. The fatty acids from linseed oil (1,000 g/ml after a three-hour incubation) resulted in 100 percent dead carcinoma cells."[10]

Flax seed oil has been used for decades as an essential part of the therapies at the renowned Gerson cancer clinic.

Several studies have found that the use of flax seed oil reduces pain, inflammation and swelling caused by arthritis.[11]

Flax seed oil has been found to lower high cholesterol and high triglyceride levels. It plays an essential role in softening and balancing the hardening effects of cholesterol in cellular membranes and helps keep veins and arteries soft and pliable.[12]

Atheletes and fitness buffs find that using flax seed oil regularly, alone and in the oil-protein combination greatly increase the oxygenation of their bodies as well as increasing their stamina and improving their recovery time.

And finally, many women find that the use of flax seed oil with a good diet and other nutritional support has often cleared up PMS symptoms as well as prevented them from developing   stretch marks after having children.

---

1. Budwig, Dr. Johanna, *Flax Oil As A True Aid Against Arthritis, Heart Infarction, Cancer and and OtherDiseases*, Vancouver, Canada: Apple Publishing, 1992.
2. Erasmus, Udo, *Fats and Oils*, Canada: Alive 1990.
3. Fischer, William L., *How to fight Cancer And Win*, Canada:Alive, 1987.
4. Rudin, Donald, M.D., and Felix, Clara, *The Omega 3 Phenomenon*, New York: Rawson Associates, 1987.
5. Budwig, op cit.
6. Wysong, Dr. Randy L., *Lipid Nutrition*, Michigan: Inquiry Press, 1990.
7. Rudin, op cit.
8. Budwig, op cit.
9. Fischer, op cit.
10. Ibid.
11 .Budwig, op cit.
12. Ibid

5

How Much To Use?

This is a very tricky question, and probably the last thing we want to hear is yet another edict from the ivory tower telling us how to live our lives.

Yet, we know we have to do something, because when three-quarters of the population is dying from diseases (heart disease and cancer) that hardly existed one hundred years ago, it is obvious that we are doing something wrong. The difficulty in finding a clear answer is that there has been such a thorough breakdown in our social structure and the memory of our cultural wisdom, as well as such a radical change in the production and preparation of our foods, that most of us no longer have a cultural heritage to turn to for answers and direction.

For myself, I have sought answers to these questions from four sources. The first is my own personal experience – seeing what nourishes and heals me. Second, listening to my friends and clients; third, objectively examining the hardcore scientific research; and fourth, looking at the studies of societies of people who live long, healthy lives free of most diseases.

In looking at the diet and nutrition of traditional, healthy peoples, some very clear answers come to light. First, there are no records of healthy societies where people lived exclusively on brown rice and vegetables, carrot juice, or wheatgrass juice and sprouts; nor for that matter did they consume hot dogs, french fries, milk shakes, ice cream, TV dinners, Coca Cola or other such creations of the modern chemical laboratories. While there have been a few healthy vegetarian cultures, most of them have not been strictly vegetarian. Instead, traditionally their diets have been broader based, using good amounts of dairy products, eggs, fish, fowl and some using meat.[1,2]

Healthy cultures were well nourished, most ate plenty of vegetables, and some fruit; some used grains or a form of tuber; and nearly all had good sources of high quality fat and protein in the form of beans, seafood, dairy foods, wild game or fowl and livestock. It is also important to note that all their food was either from wild sources or else organically grown.[3]

What no healthy society used were processed oils, refined sugar, refined and bleached flour products, caged animals, synthetic preservatives and food colorings, pesticides and herbicides, synthetic, acidulated fertilizers, chlorine in the water supply, or a host of other poisonous compounds that exist today in our food, air, and water supply.

The other common denominator of healthy peoples is that they all had a lot of physical work and activity – fishing, farming, hunting, building, washing, cooking, and walking.

One thing that becomes clear when one studies the research is that, while there are general guidelines, there is also a great deal of individual variation – some call it biochemical individuality. Some people have a much higher or different metabolism than others, or a much more vigorous physical work or lifestyle, and their bodies will need several times as much fat as others. Others, through genetic heritage or illness may only tolerate small amounts of very high quality oils in their diets.

As a general guideline, most people need anywhere from 15% to 30% of their calories from a combination of saturated and unsaturated fats – which is from 2-1/2 tablespoons to 5 tablespoons a day. This includes salad oils, cooking oils and fats found in foods including beans, nuts and seeds, meat, fish, and poultry. And of this amount, at least half should be Omega 6 fatty acids and one-quarter Omega 3 fatty acids.[4,5]

In usable terms, this means about 2 tablespoons of sunflower or similar type of oil and a tablespoon of flaxseed oil. That, combined with a little olive or hazelnut oil, or butter, will yield the minimum daily requirements for essential and non-essential fatty acids.

Again, this is a minimum. Some people will function optimally on substantially higher amounts of fats, but they should be good quality, unprocessed oils.

People with heart disease, cancer or similar illnesses partially caused by a lifetime of poor dietary habits of consuming excessive amounts of saturated, hydrogenated and trans fats need to be very careful to eliminate all forms of the aforementioned fats and all processed oils and to use a larger amount of flax seed oil – 1 to 2 tablespoons – for the first few months until their Omega 3 deficiency is restored to normal.

There are five changes that need to be made in fat consumption in the average dietary habits in the western world in order to eliminate the toxins from dietary fats, correct deficiencies and imbalances, and make real

steps towards restoring good health:
1) Reduce overall fat intake
2) Eliminate processed oils from diet
3) Reduce saturated fat intake
4) Use correct amount of Omega 3 fatty acids
5) Use correct amount of Omega 6 fatty acids

---

1. Schmid, Ronald, *Traditional Foods Are Your Best Medicine,* New York, Ballantine, 1987.
2. Price, Weston, *Nutrition and Physical Degeneration,* La Mesa, California, Price-Pottenger Nutrition Foundation, 1945.
3. Schmid, op cit.
4. Rudin, Donald, M.D., and Felix, Clara, *The Omega 3 Phenomenon,* New York, Rawson associates, 1987.
5. Galli, Claudio, and Artemis P. Simopoulos, *Dietary Omega 3 and Omega 6 Fatty Acids: Biological Effects and Nutritional Essentiality,* New York and London: Plenum Press, 1988.

6

Why Organic?

Use of independently certified organic seed in the production of unrefined oils is critical for several reasons:

First, the composition and quantity of key nutrients in organic seeds is much higher than in nonorganic seeds. This has been established by many scientific studies.[1,2,3,4]

Second, nonorganic seed can contain high levels of pesticides and herbicides.[5,6,7] With the great increase in cancer and immune system disorders due in a large part to our excessive exposure to toxins, it is a keen concern of every intelligent person today to reduce their intake of carcinogenic chemicals wherever possible. The amounts we are exposed to daily in our air, water, and food that are beyond our control are far in excess of what our bodies can tolerate and still remain in good health.

Third, this is an important step an individual can take to help the environment. Buying independently certified organic foods supports the conversion to organic farming methods.

A significant problem in the natural and organic food movement today is that many companies are trying to capitalize on the growing public interest in organic foods by falsely marketing their commercially-produced products as organic, or simply giving their own corporate guarantee which means nothing.

Recently, one oil producing company lost its FVO status due to misrepresentation of their oil as being made from organic flax seed, when in reality they were using commercially-produced seeds. Another company has also been claiming to produce organic oils by presenting falsely obtained documents and using receipts for organic seed while using commercial seed for the oil they produce.

These situations and others show the crucial importance of sincere producers of organic food products being certified by independent companies like FVO and OCIA that adhere to internationally recognized standards in the organic foods industry.

Many companies are trying to fool the public by certifying their own products as organic when in reality they are not. This business of companies certifying themselves is not credible and is basically meaningless to the informed, discerning consumer. Can you imagine if the FAA (Federal Aviation Administration) certification was not mandatory and companies like American Airlines and TWA could certify their own safety standards?

When you see the words organic or certified organic on the label, the question to ask is, certified by whom? According to what standards? As far as government regulations go (especially on the enforcement level), coal tar can be put in a bottle and labeled certified organic, although this is changing.

It is for this exact reason that honest, conscientious organic farmers and producers of organic foods submit their growing and producing methods to review by reputable independent certification companies like FVO (Farm Verified Organic) or OCIA (Organic Crop Improvement Association).

Knowledgeable consumers who buy products that are independently certified support the organic movement and the development of quality, nutritious foods and help improve our environment as well.

---

1. Jensen, Bernard, and Anderson, Mark, *Empty Harvest*, New York: Avery Publishing, 1990
2. Schmid, Ronald, *Traditional Foods Are Your Best Medicine*, New York: Ballantine, 1987.
3. Hamaker, John D., *The Survival of Civilization*, California: Hamaker-Weaver, 1982.
4. Carlat, Theodore Wood, *Organically Grown Food*, California: Wood Publishing, 1990.
5. Ibid.
6. Jensen, op cit.
7. Hamaker, op cit.

Fatty Acid Profile For Fresh Pressed Organic Vegetable Oils

Oil	Saturated Fat	Omega-9 Oleic Acid	Omega-6 Linoleic Acid	Omega-3 Alpha-Linolenic Acid
Almond	9%	65%	26%	
Canola	6%	60%	24%	10%
Flax	9%	16%	18%	57%
Hazelnut	7%	76%	15%	
Olive	10%	82%	8%	
Pistachio	12%	54%	31%	
Pumpkin	9%	34%	42%	15%
Safflower	8%	13%	79%	
Sunflower	12%	19%	69%	
Sesame	13%	46%	41%	
Walnut	16%	28%	51%	5%

- ☐ Saturated Fat
- Omega-9 Oleic Acid Monosaturated
- Omega-6 Linoleic Acid Polysaturated
- Omega-3 Alpha-Linolenic Acid Polyunsaturated

Culinary Uses of Fresh Pressed Organic Vegetable Oils

Type of Oils	*Prepared Foods	Salads	Baking	Light Sauteing
Almond	*	*	*	*
Canola	*	*		
Flax	*	*		
Hazelnut	*	*	*	*
Olive	*	*	*	*
Pistachio	*	*		
Pumpkin	*	*		
Safflower	*	*	*	
Sunflower	*	*	*	
Sesame	*	*	*	*
Walnut	*	*		

We do not recommend high temperature frying or deep frying for healthful use of oils.

*Prepared Foods - in this category we are indicating to add these oils <u>after</u> the food has been cooked.

© 1992 John Finnegan

7

Summary

It is not the position of this book that nutritional deficiencies are the sole cause of modern diseases. Obviously, there are several other factors, and all have essential parts to play in health and illness. Exercise, other essential nutrients, pathogenic organisms, environmental poisons and poisons in foods, social stresses, right livelihood, inner peace, self love, honesty, and responsibility in living all have critical roles. This book explores just one aspect of the entire picture.

Many researchers have found that today's diets are seriously deficient in key vital nutrients. These nutritional deficiencies of vitamins, minerals, enzymes, proteins, and fatty acids are a major cause of most current illness.

A vital concern in any nutritional program is to include adequate amounts of the essential Omega 3 and Omega 6 fatty acids as well as to eliminate any sources of oils containing poisonous trans fatty acids, hydrogenated fats, free radicals, rancid fats, pesticide and herbicide residues, solvent residues, and other toxic substances. This includes eliminating all commercial vegetable oils and almost all health food store oils.

Of all the foods that we consume, none is as severely processed and converted into poisonous substances as are the fats and oils. Use of high temperatures and chemical solvents, as well as exposure to light and oxygen, in the processing methods of nearly all oils produced today, destroys much of the essential Omega 3 and 6 fatty acids, and creates rancidity, poisonous trans fatty acids and many other toxic compounds.[1,2,3]

Essential fatty acids belong to two groups, which scientists have named the Omega 3 and the Omega 6 families of essential fatty acids. Since both are essential for good health, we must obtain these two fatty acids from what we eat.

Studies have found that the foremost nutrient that most of us are deficient in is that of the key Omega 3 fatty acid. An inadequate intake of this vital nutrient has been clearly linked with helping cause most modern diseases. Today, because of food processing, the average diet contains only 1/6th of what is needed and what the average diet in 1820 contained (and in many cases 1/20th to 1/100th).[4]

Just as an excess of cholesterol with an insufficient amount of Omega 3's can cause imbalance and problems, so too can an excess amount of Omega 3 fatty acids. Some Eskimos develop a condition in which their cellular membranes become so permeable that the fluid leaks out, and the slightest abrasion will cause severe swelling and leakage of fluid into tissues. My observation is that this occurs when a person ingests an excess of Omega 3's. The correct dosage of flax seed oil has been found to be around 1 and 2 tablespoons daily as a maintenance dose. This is a general amount, and one should consider factors like age, size and weight of the individual, dietary history, present diet, quantity of cholesterol normally consumed, quantity of cold water fish normally consumed, etc.

There are only two main sources of Omega 3 fats: fish oils and organic flax seed oil. Flax seed oil is the richer source of Omega 3 fats, requires less processing, tastes better, contains no toxic substances, is more stable, and is less expensive.

The Omega 6 fatty acids have also been found to be of critical importance in maintaining good health. Most of us eat ample (if not excessive) amounts of Omega 6 fatty acids, in our consumption of sunflower, safflower and other oils as well as from eating nuts and seeds, but insufficient amounts of Omega 3's. Omega 6's must be taken with an adequate amount of Omega 3's or else they can have a deleterious effect on health. An acceptable ratio among researchers is about 4 to 6 parts Omega 6 to 1 part Omega 3 in the diet.[5,6]

The Omega 9 fatty acid, mostly found in olive, hazelnut, sesame, canola and almond oil has long been valued for its beneficial effect on liver and gall bladder function, and its use is recommended to help prevent heart disease.

GLA, gamma-linolenic acid, is being used to supply the body with essential raw material with which to build vital prostaglandins. Borage seed oil is nature's highest source of gamma-linolenic acid, more than twice as high as evening primrose oil. Sometimes the best use of GLA is in small amounts for a short duration of time (one to three months) with plenty of supportive Omega 3's. Often the body willl regain its ability to produce GLA directly from the Omega 6 fatty acids. Also, some people begin to develop problems when taking GLA indefinitely without adequate Omega 3 fatty acids.

To insure that we do not burden our bodies with toxic substances and that we receive adequate intake of our essential fatty acids, we should use only high quality oils with extra flax seed oil and/or include regular amounts of cold water fatty fish in our diets.

Many companies today are producing poisonous oils by cheap processing methods and misleading the public into thinking they are cold-pressed by putting the oils into pretty bottles and selling them to the health food stores.

High quality oils need to be expeller pressed at temperatures below 118 degrees F. instead of solvent extracted. They should not be subjected to high heat temperatures, to deodorizing, bleaching, alkali refining, or winterizing processes. They need to be produced by light and oxygen excluding methods, bottled in containers that prevent the further exposure to light causing rancidity, and where possible they should be produced from third party certified organically grown seed.[7]

At this time, I have found only four companies producing oils meeting these standards of quality: Flora Oils, Galaxy Enterprises, and Omega Nutrition in Canada, and Flora, Omega Nutrition/Arrowhead Mills, and Seymour Organic Foods in the United States. Omega Nutrition/Arrowhead Mills, and Galaxy Enterprises, Ltd. have the added advantage that the oils are bottled in completely light-excluding containers. Omega Nutrition and Arrowhead Mills are the only companies that have independent third party certification (FVO and OCIA) that they use exclusively organically grown nuts and seeds.

These oils are available for purchase or can be ordered from your local health food or grocery store.

1. Mensink and Katan, "Trans Fatty Acids and Lipoprotein Levels" *New England Journal of Medicine*, Vol. 323, NO.7 August16,1990.
2. Swern, Daniel, Editor, *Bailey's Industrial Oil and Fat Products*, New York: John Wiley and Sons, 1979.
3. Gittleman, Ann Louise, M.A., *Beyond Pritican*, New York: Bantam, 1989.
4. Rudin, Donald, M.D., and Felix, Clara, *The Omega 3 Phenomenon*, New York: Rawson Associates, 1987.
5. Galli, Claudio, and Artemis P. Simopoulos, *Dietary Omega 3 and Omega 6 Fatty Acids: Biological Effects and Nutritional Essentiality*, New York and London: Plenum Press, 1988.
6. Rudin, op cit.
7. Carlat, Theodore Wood, *Organically Grown Food*, California: Wood Publishing, 1990.

Bibliography

Bates, Charles, Ph.D., *Essential Fatty Acids and Immunity In Mental Health,* Washington: Life Sciences Press, 1987.

Budwig, Dr. Johanna, *Flax Oil As A True Aid Against Arthritis, Heart Infarction, Cancer and Other Diseasess,* Vancouver, Canada: Apple Publishing, 1992.

Carlat, Theodore Wood, *Organically Grown Food,* California: Wood Publishing, 1990.

Carlson, D.J., Supruchuck, T., and Wiles, D.M., "Photo-oxidation of Unsaturated Oils: Efects of Singlet Oxygen Quenchers,: Division of Chemistry, Nat. Res. Conl. of Canada: J.O.A.C.S. Vol. 53, October 1976, p. 656-9.

East/West Journal, Februrary 1988, "Butter vs. Oil" Rudolph Ballantine, M.D.

Erasmus, Udo, *Fats and Oils,* Canada: Alive, 1986.

Faria, J.A.F., de Viscosa, U., and Mukai, M.K., "Use of a Gas Chromatographic Reactor to study Lipid Photo-Oxidation," Rutgers University, New Jersey: J.O.A.C.S.
Vol. 60, No. 1, 1983.

Finnegan, John, *Fats and Oils: Promise Or Poison, the inside story about oils that nourish us and those that poison - and the new breakthrough methods of producing good oils,* California: Elysian Arts, 1992.

Finnegan, John, and Gray, Daphne, *Recovery From Addiction, a comprehensive understanding of substance abuse - with nutritional therapeies for recovering addicts and co-dependents*, California: Celestial Arts, 1990.

Finnegan, John, and Cituk, Kathy, *Natural Foods and Good Cooking*, California: Elysian Arts, 1989.

Fischer, William L., *How To Fight Cancer And Win*, Canada: Alive, 1987.

Galland, Leo, M.D., *Superimmunity For Kids*, New York: Delta, 1988.

Gittleman, Ann Louise, *Beyond Pritikin*, New York: Bantam, 1988.

Gunstone, Harwood and Padly, *The Lipid Handbook*, London: Chapman and Hall, 1986.

Hamaker, John D., *The Survival of Civilization*, California: Hamaker-Weaver, 1982.

Jensen, Bernard, and Anderson, Mark, *Empty Harvest*, New York: Avery, 1990.

Johnston, Ingeborg M., C.N., and Johnston, James R., Ph.D., *Flax Seed Oil And The Power of Omega 3*, Connecticut: Keats Publishing, 1990.

Journal of American Oil Chemists Society, Vol. 61, Mar. 1984, Flavor and Oxidative Stability of Hydrogenated and Unhydrogenated Soybean Oils. Efficacy of Plastic Packaging. K. Warner, T.L. Mounts, N.R.R.C., Agr. Research Ser., USDA.

Journal of American Oil Chemists Society, Vol. 62, Aug. 1985, Study of Oxidation by Chemiluminescence. IV. Detection of Levels of Lipid Hydroperoxides by Chemiluminescence. Y. Yamamoto, E. Niki, R. Tanimura, Y. Kamiya, Dept. Reac. Chem Fac. Engr., U. of Tokoyo, Japan.

Mensink and Katan, "Trans Fatty Acids and Lipoprotein Levels" *New England Journal of Medicine*, Vol 323, No. 7, August 16, 1990.

Price, Weston, *Nutrition and Physical Degeneration*, La Mesa, California: Price-Pottenger Nutrition Foundation, 1954.

Rudin, Donald, M.D., and Felix, Clara, *The Omega 3 Phenomenon*, New York: Rawson Associates, 1987.

Schmid, Ronald, *Traditional Foods Are Your Best Medicine*, New York: Ballantine, 1987.

Stier, Bernard, *Secretes des Huiles de Premiere Pression A Froid*, Canada: 1990.

Swern, Daniel, Editor, *Bailey's Industrial Oil and Fat Products*, New York: John Wiley and Sons, 1979.

Wysong, Dr. Randy, *Lipid Nutrition*, Michigan: Inquiry Press, 1990.

A Note to Companies
Not Mentioned In This Book

The Facts About Fats is an excerpt from a much larger and more extensive book, *Fats and Oils: Promise or Poison, the inside story about oils that nourish us and those that poison - and the new break-through methods of producing good oils,* by John Finnegan, to be published by Elysian Arts in 1992.

Any companies not mentioned in this book that produce truly fresh-pressed oils at temperatures below 120 degrees F., in a light and oxygen excluding environment, pressed from third party certified organically grown seed, and bottled in light-excluding, inert gas-sealed containers, please contact me at the following address so I can include you in future editions of my books.

Sincerely,
John Finnegan

Elysian Arts
29169 Heathercliff Road
Suite 216-428
Malibu, CA 90265

About The Author

John Finnegan, was born in Greenwich Village and raised in Long Island, the jungles of Latin America, and the beaches and redwoods of Northern California. He began writing his first book when he was nine years old - the story of his family's journey from New York to Lima Peru. They were the first people to drive the length of Central America, often having to cut their own road through the jungle with machetes, shovels, and pick-axes.

At nineteen, he began to research the biochemical basis of physical and mental illness, which included studying with many of this century's leading medical pioneers. He studied life sciences and social sciences at San Francisco State University, College of Marin, and continued his studies with Dr. John Christopher, Dr. B. Sahni, Dr. Broda Barnes, Wendell Hoffman, Piro Caro, other holistic researchers, and in several medical centers. John Finnegan is the author of 8 books, including *Recovery From Addiction*, which he co-authored with Daphne Gray, and is published by Celestial Arts. He lectures and conducts seminars, and gave presentations at both the 1987 and 1988 San Francisco Whole Life Expos.

Recovery
From Addiction

A Comprehensive Understanding
of Substance Abuse
with Nutritional Therapies
for Recovering Addicts & Co-Dependents
by John Finnegan and Daphne Gray
248 pages $9.95

"**Recovery From Addiction** creates a paradigm of health for the complex subject of drug abuse, rather than treating it as just another disease. I feel this is the only way to approach healing addictions on a human level."

- Stephen Langer, M.D.
Author, **Solved: The
Riddle of Illness**

This is the first book that discusses all the factors involved in creating and healing the addictive or alcoholic condition.

Recovery From Addiction explores the social , economic, and political issues, the psychological and spiritual dimensions, and the problem of codependency. It also provides the most up-to-date, complete information on the underlying metabolic causes of addiction and the invaluable nutritional therapies that can help to corect them.

And most of all **Recovery From Addiction** provides solutions...

An *Elysian Arts* book published by *Celestisal Arts*.
Available wherever books are sold.

Amazake

Naturally Delicious & Nutritious
Rice Beverage
By **John Finnegan** and **Kathy Cituk**
80 pages $6.95

"The best way to create a strong, healthy body and mind is to live well and eat wholesome, unrefined foods.... Amazake, made from organic whole grain rice, provides the nourishment we need without the dangers of preservatives or added sugars, making it a good nutritional choice."

- From the book

Amazake is the first book to be written for the general public about this unique, delicious, all natural drink. Amazake is made by mixing whole grain brown rice with koji (a rice culture) or enzymes, and then incubating it until the complex starches are broken down into easily digestible, natural sugars. Because it is made simply with only a few ingredients, amazake is naturally sweet, and virtually nonallergic.

This book discusses the nutritional benefits and versatility of amazake and explains how it can be used as a dessert, snack, natural sweetening agent, baby food, or made into kefir-yogurt, smoothies, salad dressings and ice cream.

An *Elysian Arts* book published by *Celestial Arts*
Available wherever books are sold.

The Facts About Fats
A Consumer's Guide to Good Oils
by John Finnegan
64 pages $5.95

This little gem of a book gives a clear and concise understanding of the critical role that fats and oils play in promoting good health. It tells the truth about the so-called cold-pressed health food oils and shows how good oils are really made. Written in a lively, honest, hard-hitting style, it exposes margarine and processed oils as major contributors to heart disease, cancer, and other modern diseases. It discusses the Omega 3 and Omega 6 essential fatty acids, the wonders of flax seed oil, and many other concerns of vital importance to all of us striving to live in our technologically altered society.

"In his new book, **The Facts About Fats**, John Finnegan has, through his dedicated study, personal observation, experience and desire for integrity, assembled this critical body of nutritional information about fats and oils.

This book answers questions about fats and oils that have occurred to every serious student of nutrition, and is essential reading for those genuinely concerned with having a healthy diet for themselves and their families."

-- Brian Roettger, D.C.

The Facts About Fats
A Consumer's Guide to Good Oils
by John Finnegan

Send _____ *The Facts About Fats* @ $ _____ = $ _____

1to 2 cpies......................	$5.95
3 to 9 copies..................	$5.00
10 to 49 copies.............	$4.25
50 to 99 copies.............	$3.50
Over 100 copies	$3.00

Book Total = $ _____

Shipping & Handling*

1 to 2 items	$3.00
3 to 9 items	$5.00
10 to 49 items	$7.00
50 to 99 items	$12.00
100 items	$20.00

Sales Tax = $ _____
8.25%(California Residents Only)

*Shipping = $ _____

Total = $ _____

*Please Note: Shipping outside of continental U.S., including Hawaii, Alaska, and Canada is double. All books are shipped either by First Class Mail or UPS.

Name _____

Address _____

City _____

State/Prov _____ Zip/PC _____

Please make checks payable
(in U.S. funds only) to

Elysian Arts
29169 Heathercliff Road
Suite 216-428
Malibu, CA 90265